Animals in Disguise

by Andrew Einspruch

EDUCATORS PUBLISHING SERVICE
Cambridge and Toronto

Series Authors: Kay Kovalevs and Alison Dewsbury
Series Consultant: Melinda Rice
Commissioning Editors: Sarah Russell, Bonnie Lass,
and Sethany Rancier Alongi
Senior Editorial Manager: Sheila Neylon
Text by Andrew Einspruch
Edited by Sally Paxton and Elissa Gershowitz
Designed by Anna Cesaro, Andrea Jaretzki, and Maria Vallianos

Making Connections program developed by Educators Publishing Service,
a division of Delta Education, LLC and by Harcourt Education, a division of
Reed International Books Australia Pty Ltd

ISBN 0-8388-3326-8

2010 2009 2008 2007 2006
10 9 8 7 6 5 4 3 2 1

Printed in China

Acknowledgments
The author and publisher would like to thank the following for permission to reproduce the copyright material in this book.

Photographs
Australian Picture Library/ Peter Johnson, p. 4; Australian Picture Library/ David Muench, p. 6; Australian Picture Library/ Albrecht G. Schaefer, p. 5; Paul & Joyce Berquist/ Animals Animals, p. 7; Jeff Jeffords/divegallery.com, pp. 14–15; Dr. Kenneth L. Krysko, p. 11; naturepl.com/ John Cancalosi, p. 10; naturepl.com/ Peter Oxford p. 20; photolibrary.com, cover, pp. 1, 16, 18–19, 21–22; Ribbit Photography, p. 8; Seapics, pp. 12–13.

Contents

A Matter of Survival

If you were an animal in the wild, staying alive would be more important to you than anything else. To survive you need to have enough to eat. You also need to make sure you don't get eaten!

Animals use many tricks to help them survive. Some run fast. Others climb or jump. Many animals use **camouflage**. This means that the way they look acts as a disguise. This disguise helps animals to blend in with their surroundings so they are hard to see.

If you are a predator, blending in helps you get close to the animals you hunt without being noticed.

This leopard blends in with its surroundings so it can sneak up on its prey.

If you are the animal being hunted, or the prey, blending in makes it hard for the hunters to find you.

Can you see the gecko? It looks like it is part of the tree.

Hidden in the Desert

Deserts are harsh, dry places to live. The rain usually comes in just a few downpours each year. Deserts are usually very hot during the day, and very cold at night.

To live in a desert, you have to be tough. Desert animals do not need much water. They have to be clever about finding food. Most of them need to be good at hiding. Many desert animals are sand-colored. This helps them blend into the sand and rocks.

The color of the bobcat's fur helps it to hide on sandy, rocky hillsides.

You Can't See Me

Horned lizards are very tough desert animals. They have spikes, or "horns," that poke out of their head and the sides of their bodies. These horns make it hard for snakes to swallow them. Some horned lizards protect themselves by shooting blood from small openings in the corners of their eyes.

Horned lizards have small, short legs. When they move, they are slow and awkward. So how do horned lizards stay safe from hawks and roadrunners? When they sense danger, the horned lizards do not run. Instead they keep very still. Their light-brown and gray bodies can be hard to see in the sandy and gravelly places where they live.

This type of camouflage is called **cryptic coloration**. This means that the patterns and colors of the body match the animal's surroundings.

The horned lizard protects itself by keeping very still and blending in with its surroundings.

Desert Danger

Western coral snakes also live in the desert. Instead of trying to hide, they use their bright colors to scare away any predators. Western coral snakes have wide red and black bands, as well as narrow yellow and white rings. These colors mean, "Watch out, I'm dangerous!" When western coral snakes bite, their venom is twice as powerful as a rattlesnake's poison.

The western coral snake hides under rocks or in the dirt. The snake stays cool during the day and comes out to hunt at night.

Snake Fakes

How do you survive if you're not a dangerous snake? One way is to pretend you are dangerous.

That's exactly what the scarlet kingsnake does. Like the coral snake, it has red, yellow, and black bands on its body that seem to say, "Warning! Warning!" The difference is that scarlet kingsnakes are harmless.

This type of camouflage is called **mimicry**. The animals mimic, or imitate, a different animal to fool their predators.

Is this scarlet kingsnake dangerous or harmless? Would you be fooled?

Hidden in the Ocean

Like land animals, many sea animals use camouflage to help them survive. Take the carpet shark. Unlike other sharks, the carpet shark is not a fierce hunter. It can hide from predators because it looks like a carpet on the bottom of the sea.

One type of carpet shark is the tasseled wobbegong. Its head is hidden by a shaggy fringe of skin flaps around its mouth. The fringe looks like the long strings of a carpet. The fringes on the tasseled wobbegong hide the shark. They also attract other fish, making the fish easier for the shark to catch.

The tasseled wobbegong blends in with the patterns and shadows on the bottom of the ocean.

The tasseled wobbegong's body is covered with spots, bands, and other marks that help the shark to hide in the coral reefs. This type of camouflage is called **disruptive coloration**. The markings look confusing to other animals. Tigers use this same trick, as they move through forests or tall grass.

Decorations that Deceive

The decorator crab uses a special trick to remain safe from predators. The crab decorates its body using things it finds on the ocean floor! This confuses the fish, seals, and otters that prey on the decorator crab.

Decorator crabs are not picky about what they use for decoration. Depending on where they live, they might attach algae, seaweed, or small rocks. Sometimes they even use animals like sponges, corals, and sea anemones [**say** *a-NEM-o-nees*]. Anemones are a good choice because they have tentacles that can sting predators.

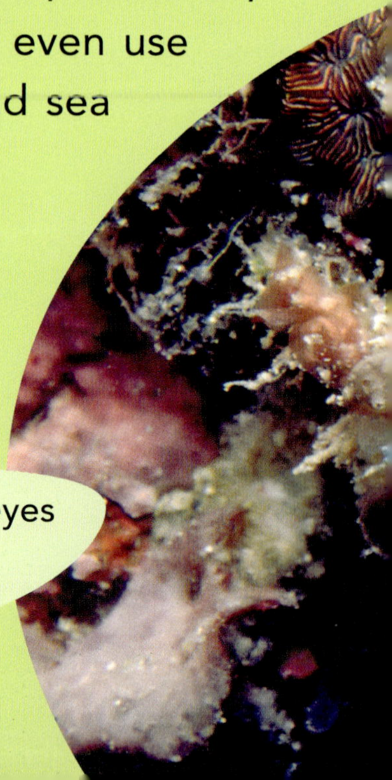

Can you spot the crab's eyes among its decorations?

How do these decorations stick? The outside of the crab is covered in small, hooked hairs or bristles. These hooks grab onto the decorations and hold them in place.

In time, decorator crabs grow big enough to defend themselves. They no longer need to decorate their shells. But plants and animals can still get caught in the bristles on their backs and continue to grow there.

Plant or Animal?

Look at this picture. Are you looking at a plant or an animal? If you said "animal," you are right. This is a leafy sea dragon, a close relative of the sea horse.

How do sea dragons survive in the ocean? They trick animals into thinking they are plants. They have skin growing out of their bodies and fins that look like bits of seaweed.

When sea dragons swim, they rock in a rhythmic motion. This makes them look like seaweed floating in the water. Sea dragons do not move very fast so they need to be very good at hiding.

Sea dragons also use a type of camouflage called **color changing**. They can change color to match their surroundings. If they move to a different colored plant, they can change their skin to blend in better. Their color is also affected by what they eat.

Hidden at the North Pole

The Arctic, like the desert, is a harsh place to live. It is freezing cold for months at a time. In winter, there is little to eat and long nights. Animals in the Arctic work hard to survive.

Arctic foxes hunt small animals like voles and lemmings. They also follow polar bears to pick up scraps from the animals that the bears kill. But they have to be careful because polar bears prey on Arctic foxes.

During summer, Arctic foxes have a dark coat that matches the dirt where they live. In winter, their coats turn white so they can hide in the snow. Some Arctic foxes live in snow all year round, so they stay white all the time.

There are other Arctic animals that turn white in winter. Some of these animals include Arctic wolves, snowy owls, long-tailed weasels, ermines, and collared lemmings.

Arctic foxes have dark fur in summer and white fur in winter.

Hidden at the South Pole

At the other end of the world, conditions are even harsher than at the Arctic. The Antarctic is the coldest, driest, windiest place on earth. One animal that survives there is the penguin.

Penguins use a type of camouflage called **countershading**. This means that one side of the penguin's body is darker than the other. If you were a bird looking at swimming penguins from above, their dark backs would make them hard to see against the dark ocean floor. If you were a shark looking at them from below, their white stomachs would blend into the light above.

Like all forms of camouflage, countershading helps the penguin hunt its prey, which includes fish and squid. It also helps the penguin avoid being eaten by its predators. Like all animals who use camouflage, for penguins, it is a matter of survival.

Further Reading

Arnosky, Jim, 1995,
I See Animals Hiding,
New York: Scholastic.

Perry, Phyllis J., 1997,
Hide and Seek: Creatures in Camouflage,
New York: Franklin Watts.

Powzyk, Joyce, 1990,
Animal Camouflage: A Closer Look,
New York: Bradbury Press.

Glossary

camouflage using the way you look as a disguise

color changing a form of camouflage where an animal changes color to match its surroundings

countershading a form of camouflage where one side of an animal's body is darker or lighter than the other

cryptic coloration a form of camouflage where an animal's body matches its surroundings

disruptive coloration a form of camouflage where spots, stripes, and markings on the animal confuse the eye and make it hard to pick out the animal against its background

mimicry when an animal imitates another animal

Index